FLORIDA'S
Unsung Wilderness

The Swamps

PHOTOGRAPHY AND TEXT BY
CONNIE BRANSILVER AND LARRY W. RICHARDSON

FOREWORD BY JANE GOODALL, PH.D.
PREFACE BY STUART D. STRAHL, PH.D., AUDUBON OF FLORIDA

WESTCLIFFE PUBLISHERS

www.westcliffepublishers.com

ACKNOWLEDGEMENTS

Thanks first to Aldo Leopold, who went before us and said it all: "There are some who can live without wild things, and some who cannot." We cannot. Edward Abbey in *The Journey Home* said, "Any scientist worth listening to must be something of a poet, must possess the ability to communicate to the rest of us his sense of love and wonder at what his work discovers." They both placed into perspective how we felt about the natural world, these places in the swamp, and our yearning to illustrate it with prose and photography so others might follow with a sound conservation ethic.

We thank Jane Goodall, our friend, our inspiration, and the tireless messenger giving all of us reason for hope.

Many more encouraged us and gave actual assistance. All the staff at the Florida Panther National Wildlife Refuge, at Fakahatchee Strand State Preserve, at Corkscrew Swamp Sanctuary, at Big Cypress National Preserve, and on private lands. We interviewed many, only some of whom are quoted in "The Good Old Days," who gave freely of their time and experience. The staff at Collier Mosquito Control District and alligator experts Brady and Mei Lin Sanchez Barr and Mike Cherkiss gave us essential technical advice and assistance. And the folks at the Wildlife Rehabilitation Clinic at the Conservancy of Southwest Florida let us release many critters back to the wild, including Jane Goodall's alligators and our vixen.

It is our families who made this book possible — Melissa, Lauren, Wendy, Holly, Lea, and Ed. We thank you.

Most of all, we thank the unsung heroes working with wildlife and helping to protect our environmental heritage so we can pass it on to our children and their children with pride.

Connie Bransilver and Larry W. Richardson

ISBN: 1-56579-386-2

Text and Photography Copyright:
Connie Bransilver and Larry W. Richardson, 2000. All rights reserved.
LarCon Productions   www.LarCon.com

Editor: Dianne J. Nelson
Designer: Dianne J. Nelson
Production Manager: Craig Keyzer

Published by:
Westcliffe Publishers, Inc.
P.O. Box 1261
Englewood, CO 80150
www.westcliffepublishers.com

Printed in Hong Kong by H & Y Printing, Ltd.

Library of Congress Cataloging-in-Publication Data
Bransilver, Connie, 1942-
    Florida's unsung wilderness : the swamps / photography and text by Connie Bransilver and Larry W. Richardson ; foreword by Jane Goodall.
        p. cm.
    ISBN 1-56579-386-2
    1. Swamps--Florida--Big Cypress Swamp Region. 2. Swamp ecology--Florida--Big Cypress Swamp Region. I. Richardson, Larry W., 1954- II. Title.

    QH105.F6 B73 2000
    577.68'09759'44--dc21
                                                                        00-043260

For more information about other fine books and calendars from Westcliffe Publishers, please contact your local bookstore, call us at 1-800-523-3692, write for our free color catalog, or visit us on the Web at www.westcliffepublishers.com.

[OVERLEAF, PAGE 1]
Water Road in the Corkscrew Regional Ecosystem Watershed, adjoining Corkscrew Marsh.

[OVERLEAF, PAGE 2]
Below the surface, the water is cool, fresh, and clear and teeming with life. The swamp IS water.

PLEASE NOTE: *All panthers captive except page 42 and kittens on page 43.*

Contents

—

Foreword by Jane Goodall   7

Preface by Stuart Strahl   13

Introduction   17

Visual Splendors   23

Where Are We?   33

The Sentinel   41

Keystones and Harbingers: Gators and Skeeters   47

Vixen   57

Mystery of the Bee-Swarm Orchid   63

Who Flies There?   69

Fire in the Garden   77

Good Old Days   83

Afterword   88

Brighter, larger-than-life flowers command attention, draw attention away even from a sunset. We can allow ourselves to be drawn in, to confront flowers in a timeless zone, and say thank you for being. The swamp lily.

Foreword

—

By Jane Goodall, Ph.D.

"That night, as I awaited sleep, I thought about the swamp.
I imagined paddling quietly, gliding across the
dark, secret water for mile upon mile, moving
through a world that once must have seemed
almost limitless — a world in which humans could
so easily become lost and disappear utterly."

Many years ago, when I was a child, I read an article about the Florida swamps. The author (I do not remember whom it was) captured my imagination as he described the mysterious scenery with its myriad plant and animal species, the haunting calls of unseen birds, the swirl of dark water as an alligator dived from a dead log. The writer was in a small boat, paddling silently beneath the interlaced branches of the trees. Everywhere was water, greenness, peace. Ever since reading that article, I had wanted to visit the Florida swamps, and the opportunity finally came during my 1998 spring lecture tour. It was, sadly, a short opportunity — a free afternoon (labeled "Rest") in the midst of my two-day stop in Naples. I was staying, along with my executive assistant, Mary Lewis, with our old friends, Ed and Connie Bransilver, and I had met Larry Richardson for the first time the evening before. When I told him I'd never been to the Florida swamps, he had a suggestion: What about visiting the Florida Panther National Wildlife Refuge during your free hours? What a wonderful idea.

Larry decided it would be a good opportunity to release two small alligators into the swamp from the Conservancy of Southwest Florida's Wildlife Rehabilitation Center — they were ready to return to the wild. We loaded their cages into Larry's vehicle, picked up his twelve-year-old daughter, Lauren, and set off. It did not take long to reach the refuge and, once we got there, it was amazing how

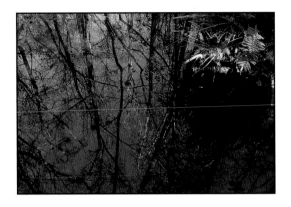

quickly civilization dropped behind us as we entered the outskirts of the swamp, how rapidly the beauty of the natural world crept into my very being. This was a different world from my own Gombe forests in Tanzania, Africa, and the flora and fauna were strange, but the sense of belonging was the same.

The two little alligators seemed reluctant to leave their human guardians at first and had to be encouraged to abandon their cages. But eventually both entered the water and slowly glided away to freedom. Larry guided us to the range of a pair of barred owls, and we were fortunate to see them and hear their strange, haunting calls. I have always loved owls, with their silent flight and wonderfully soft feathers and those extraordinary round faces with the great luminous eyes. I lay on my back on the dry ground, and looked up through the branches and leaves to the sky — as I love to do at Gombe.

I asked Larry about the status of the highly endangered Florida panther — his special interest. Larry had seen panthers close to where we were, and we searched for a sign. Although we found no tracks, a panther had made a scent scrape on the edge of the road down which we traveled. It was good to know that these beautiful cats are still hanging on — and that there is always the chance of encountering the unexpected in the wilderness. In my imagination, glowing feline eyes were watching us as we made our pilgrimage into their land — the tiny bit of land we have left for them.

The time passed quickly, and soon it was time to go. That night, as I awaited sleep, I thought about the swamps as described in that first article I had read. I imagined paddling quietly through the swamps, gliding across the dark, secret water mile upon mile, moving through a world that once must have seemed almost limitless — a world in which humans could so easily become lost and disappear utterly. But now things are different. Human beings, relentlessly, have invaded the mysterious world of great swamps, draining the water for building schemes and roads, driving the panthers and other shy creatures even farther into the heartland of the swamp system, which is getting ever smaller. Countless people are moving to this part of Florida because they have a romantic love for the natural world — a love that is destroying the very things they love. Here there is still wilderness, but it is so fragile, and already it has been changed as a result of human interference.

Sometimes it is hard not to become depressed when looking around at the problems that we humans have inflicted on our planet — the overpopulation, overconsumption, destruction and fragmentation of habitats, pollution — all because of our arrogant assumption that our needs and wants must come first. So much wilderness has been destroyed, so many plant and animal species have become extinct, and so many of the rest are now endangered. What we must realize, before it is too late, is that the damage we inflict on nature so often leads to massive human problems. In Southwest Florida, for example, the fragile wetlands are nature's water filter as well as the water supply for humans and other animals. And already this precious resource has been damaged by human meddling — all in the name of progress, of course. Sometimes human actions in one area — such as interference with the flow of water, drainage of outlying swamp areas, pumping in overflow water — can affect other, more remote areas through simple chain reactions. Biologists such as Larry, by bringing this to the attention of nature lovers and lawmakers alike, are desperately trying to prevent further damage and to nurture what is left.

But conservationists must contend with materialistic greed. What do the developers care if some small, to them utterly insignificant, creature is threat-ened with extinction, or even completely destroyed, by their building schemes? All over the world, business interests are winning out over the efforts of those trying to protect the environment. And Southwest Florida is no different. Slowly, relentlessly, the swamps are changing. They are becoming smaller, losing out, bit by bit, to human greed,

[ABOVE]
A barred owl swoops across a clearing.

[LEFT]
Florida panther print, claws retracted, four toe pads prominent.

selfishness, and ignorance. So much damage is inflicted by ignorance.

That is why I am spending more and more of my time working with young people around the world, to increase their understanding of the problems of our time, to empower them to act now, and to work together for change. I talked to Larry's daughter Lauren about our program, Roots & Shoots, and told her that she could help me by starting a group at her school. What would they have to do? Three projects to make the world around them a better place — for animals, humans, and the environment we all share. Larry and Connie were enthusiastic. Later Larry shared with me a letter he had received from a young boy who was fascinated by the Florida panther. "I think about the Florida panther. I talk about the Florida panther. I want information," he said. And he went on to ask about the progress of a new underpass that was being built under a road that bisected the panther's range in an attempt to prevent the tragic loss of panthers being hit by cars.

The enthusiasm and desire to put things right that I have encountered in young people around the world — once they know the problems — are the main reasons I have hope for the future. Once they understand, they truly begin to care. And then they want to help. This is why a book like *Florida's Unsung Wilderness* is so important. The combination of Larry's knowledge and the magic of their exquisite photography, and the fact that they both care so deeply and passionately about this part of the natural world, makes this an especially important and inspired book. I hope it will help to sway the lawmakers as well as provide the general public with a greater understanding of Southwest Florida's swamps and the important role they play in both the physical health of the region and the spiritual health of everyone who lives here or visits.

As I travel around the globe, a box is my most precious possession, for it contains my symbols of hope. In it are objects that symbolize new scientific

inventions that will reduce or prevent pollution, other objects that symbolize the indomitable human spirit, and many that represent nature's amazing powers of resilience if we give her a chance and a helping hand. I have a feather from a peregrine falcon, a species brought back from the very brink of extinction. Another item is an antler from a Formosan spotted deer in Taiwan. The species became extinct in the wild, but as a result of a breeding program organized by 16 individuals representing various zoos, there are now about 60, roaming free in a huge national park.

And now in my collection is the cast of the footprint of a Florida panther. Like the falcon and the deer, this panther, too, has another chance. When I look at the footprint, I shall forever be reminded of a magic afternoon in the Florida swamps with two people whose skills and vision have created a new song. This wilderness is unsung no longer.

JANE GOODALL, PH.D.

Preface

❧

By Stuart D. Strahl, Ph.D.
President/CEO, Audubon of Florida

"We picture the diversity of life in its most primordial splendor,
with fantastic assemblages of species, from insects to ungulates.
A musty scent of decay pervades the atmosphere as flights
of gaudy birds wheel overhead, and agile predators
glide through the shadows, just out of sight."

For those of us who have spent time in natural surroundings, envisioning the tropics brings to mind indelible imagery. We picture the diversity of life in its most primordial splendor, with fantastic assemblages of species, from insects to ungulates. A musty scent of decay pervades the atmosphere as flights of gaudy birds wheel overhead, and agile predators glide through the shadows, just out of sight.

We might think of the dense rainforests of South America, the impenetrable jungles of Borneo, or the lush forests of the Congo. Few would imagine that such a place could be found in the tropics of the United States. Yet South Florida is home to some of the world's most majestic natural places. Some, such as the Everglades, are grand, sweeping panoramas of water, land and sky. Others take the form of bubbling crystal springs, teeming with wildlife and a panoply of aquatic creatures.

The swamps of the Big Cypress Basin are foremost among Florida's natural wonders. The lure of the swamp is subtle, with intricate living webs linking aquatic and terrestrial habitats, representing a mosaic of unsurpassed ecological diversity. Orchids, bromeliads, ferns, cypress, and hardwoods envelop the casual visitor with a warm green tapestry of life.

The Audubon Society has been engaged in conservation of South Florida's magnificent ecosystem for over a century. Audubon's work began in Florida with preservation of natural systems. Our "Audubon Wardens," a few brave souls camped in the wilderness, protected wading bird colonies from the depredations of avian plume hunters. The stakes were high, as millions of dollars' worth of egret plumes were being exported from Florida each year. By 1900, less than 10 percent of our state's plumed birds remained.

Guarding wildlife was dangerous work as well. Two of our Florida wardens,

[ABOVE]
In the swamp, light is sometimes quiet, and other times it insists on boisterous colors, bright red against it compliment, green. Swamp hibiscus.

⸺

[OVERLEAF]
Saw grass, overpowered by dew, bends as if praying to the dawn, like graceful strokes of Japanese calligraphy.

⸺

[INSET]
Crayfish are a mainstay of the diet of many wading birds.

Guy Bradley and Columbus MacLeod, were murdered in the line of duty. Thanks to the resulting fervent public outcry and the ongoing sacrifice and lonely work of these early pioneers of preservation, Audubon helped pass landmark federal and state legislation in the early 1900s, protecting all of America's wildlife.

Since those early days, Audubon of Florida has continued to lead the charge in preserving key regions of Florida's Southwest Coast, and in protecting its wildlife. In the mid-1950s, we led a nationwide grassroots effort to preserve the last of the great stands of virgin cypress swamps of the region. In 1954 we acquired 11,000 acres in the middle of the Corkscrew regional watershed and founded the Corkscrew Swamp Sanctuary. Today, Corkscrew's 2.5 miles of ecologically friendly boardwalk and our newly completed Blair Audubon Center are the shining jewels of our nationwide sanctuary system, providing nearly 150,000 visitors each year with nature-based experiences.

In the 1970s, Audubon's South Florida office helped establish the Big Cypress National Preserve to forever protect this outstanding example of tropical American biodiversity. Our scientists have been actively studying the Big Cypress for decades, producing volumes of information (including the authoritative work, the *The Big Cypress National Preserve*, in 1985).

Today, the swamps of Southwest Florida are again a front-line battle-ground, this time at the hands of rampant development. What remains of the natural system is home to many of the region's most endangered plants and animals. The Florida panther, wood stork, ghost orchid, cigar orchid, and hand fern are just a few of those threatened species dependent on this ecosystem.

Maintaining these habitats in their natural state is crucial to the ecological well-being of South Florida, yet the region is being converted to urban use at an alarming rate. Some say the

Naples–Sarasota coastal region is the fastest growing urban complex in America. Even as development encroaches on this magnificent habitat, few local residents are aware of what lies in their backyard.

This magnificent book offers the reader a glimpse of this fantastic place, and an invitation to explore. We at Audubon of Florida are thrilled to support this work. We hope that it will draw readers into the complex and magnificent ecological web of these pristine reaches and provide a deeper appreciation of nature's backyard wealth. Within these pages lies a clarion call to ensure the preservation of what remains of Southwest Florida's natural wonderland.

STUART D. STRAHL, PH.D.
President and CEO, Audubon of Florida

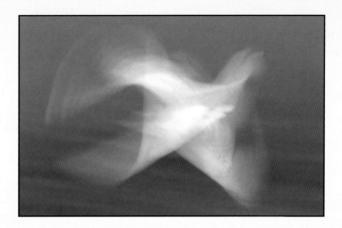

Introduction

~

"[H]ave dominion over the fish of the sea
and over the fowl of the air
and over every living thing
that moveth upon the earth."

Genesis 1:28

The oldest edict for conservation comes from the Book of Genesis, where God gives Adam dominion — *stewardship* — over the earth. While the controversy over creation and evolution flourishes, we leap over it all and face our world as it is now, looking forward to the responsibility we have to care for and nurture the earth with understanding, knowledge, and wisdom. We are but one species, albeit the most powerful. We have hope, as does Jane Goodall, that our hearts and minds can exercise dominion wisely, but we cannot love or cherish what we do not know. We introduce you to a piece of nature's wonder in all its beauty and say why it is important that we examine it. We also see how humans have lived with this sometimes hostile land, cooperated with its natural ebb and flow, and now, increasingly, diminish it. We all will lose if we fail to understand and honor its processes.

Writers such as Ralph Waldo Emerson, Henry David Thoreau, Aldo Leopold,

and, in our generation, Jane Goodall, Barry Lopez, and a noble host of others, all exalt nature's ability to comfort the soul and lift the spirit. We take refuge in the wide-open spaces, if not in actuality, then through images that resonate with our particular needs. We are refreshed by clean air, the smells of earth and trees and water flowing, vistas to thrill the eye, and wildlife that epitomizes

[ABOVE]
The swamp is ever-moist, expectant with dew, waiting to be warmed by the strength of the sun.

[OVERLEAF]
Sunset over the Big Cypress Swamp.

[INSET]
An apparition? A kite caught by the breeze? The same shape, a great egret flying, bends the mind.

[TOP]
The green anole is quick and wary.

[ABOVE LEFT]
An unknown snail, patterned by nature.

[ABOVE RIGHT]
Even the lowly fungus is an ode to the swamp's sculpture.

[LEFT]
"Tears would come to my eyes with an especially beautiful sunset."
Jane Goodall in Reason for Hope.

freedom in contrast to our own world of confinement, deadlines, machinery, and just making a living. We are lesser beings when we have not experienced wilderness in its finest forms, natural and ordered.

As photographers and writers, we reveal another dimension. We attempt to bring the essence of one piece of wilderness to all who would search these pages. *Florida's Unsung Wilderness: The Swamps* shares our vision of a place and time with others who may never have experienced the natural world we see, or it may reconfirm remembered times and experiences. From an examination of tiny wonders to broad vistas, we expose this unique environment. We document factually and graphically, but we also go beyond the obvious into realms of the surreal, with the water and trees and muted light providing the backdrop and the props. Like those who have written so we will not forget our place on earth, we, too, provide a glimpse into the innumerable processes

Florida panther

of nature. Our palette is Southwest Florida's swampland, and specifically, what we call the Big Cypress Basin. It is termed a basin because it is, in fact, like a river with no definable banks, shallow and slow moving. Visually, we want to share nature's splendid and perfect order intricately played out in this misunderstood ecosystem.

We have mostly chosen to hide the human siege on our sloughs and strands by what we have chosen to illuminate through this book. We photographed what is still good and right and turned our cameras away from assault, from plunder, from our human mistakes. There is a bigger story to tell, though. Since the beginning of the last century, at least 50 percent of our local wetlands, mostly classified as swamps, have disappeared. In the southeast quadrant of Florida, the Everglades are but a remnant of what used to be a vast, lingering river of grass with unprecedented populations of wildlife. What remains is entangled in controversial and expensive plans for restoration. In Southwest Florida, swamps have suffered, but not to the degree other habitats have. Swamps have been too formidable. Until we invented and built ever bigger, stronger machines to plunder the swamp, it has held people at bay with its hostile environment of quagmires and bugs, snakes and alligators. It has shared only morsels, and those precious morsels have been offered only to respectful suitors.

Outside the swamp proper, in South Florida, only about 10 percent of our upland pine forests still remain, the rest having gone under the 'dozer blade in the name of development. The same plight plagues scrub habitat — the sandy, sometimes dunelike environment along coastal and some inland sites dominated by scrub oak. Will our drive to develop our coastal zones and the impending sea-level rise mean its demise and the ruination of our estuaries and coastal fisheries? Red tide, once an occasional natural occurrence, flourishes

more and more often, polluting our beaches with dead sea life, fueled now by pollutants entering the Gulf of Mexico through manmade canals.

We still wonder if there will be Florida panthers in the next hundred years, or even the next decade. We know what it takes for their survival — food, water, and cover — enough space for them to hunt and breed and continue to capture our imaginations. Can we secure enough wilderness to allow their isolated populations to expand and avert extinction? People, at one time through bounty and now through systematic conversion of the wilderness, are killing off Florida's most enamored poster child. Our incessant and misdirected quest to change essential habitats for our perceived gain may halt any thought of panthers ever surviving as anything other than a zoo oddity. And with the wild panther will go all the less charismatic species protected by the umbrella of wilderness preserved.

Florida's Unsung Wilderness: The Swamps is a cry for help for places that stand unprotected from ourselves. Swamps are made up of strands — forested, slow-moving watercourses — and sloughs, the deeper sections of a strand and all the plants and animals that inhabit them. Swamps are the visual representation of aquifers — the unseen, underground source of our fresh water in Southwest Florida. Their perpetuation can be verified every time we turn on a faucet. We have tried to tell their story through snippets of text, enough to encourage you to seek more. We have told the story through pictures and captions that, we hope, will inspire as well as inform; and we have told the story as others, throughout the eons, have seen nature and the relationship of human beings with it.

So we lift the veil, reveal the swamp to you, as we see and experience it ourselves.

Visual Splendors

—

"Whoever you are, no matter how lonely, the world offers itself to
your imagination, calls to you like the wild geese, harsh and exciting —
over and over announcing your place in the family of things."

New and Selected Poems
Mary Oliver

The swamp is subtle. No huge mountain peaks grab your eye and hold it for hours or miles as you approach. No dramatic coastline, nor thundering rapids, transform water into sculptures of moving form. No dramatic fall colors roll across the breadth of vision to delight the eye. The South Florida swamp is such a cacophony of images that at first none can be differentiated.

It is a deafening assault on the senses — confusing, perhaps forbidding. Confusion might even retreat into fear of being overwhelmed with the abundance of life, unfamiliar in form and substance. At first, perhaps. But try the water. It is fresh, cool, usually clear and teeming with life — fish snapping at mosquito larvae, tiny grass shrimp — and it is flowing, ever so slowly. A blue dragonfly hovers. A barred owl calls its mate — "Who cooks for you? Who cooks for you?" — then swoops across a gap in the canopy. A pond apple plops into the

[LEFT]

In the swamp, no curtains of dramatic fall colors assault the eye, but after the first cold, measured as 50-degree lows, red maple leaves turn, piercing the green and gray tones with buttons of color.

[BOTTOM LEFT]

A dragonfly hovers — blue or golden-winged — perhaps devouring its favorite meal, a fat mosquito. Both dragonflies and damselflies have moveable heads and large, compound eyes.

[OVERLEAF]

The water is fresh, cool, clear, and teeming with life. Decaying matter joins with late-afternoon shafts of light to gild the enlarged base of a bald cypress, which may be as old as the redwoods.

[INSET]

You rarely see a live apple snail, the aquatic gastropod the size of a golf ball that seems abundant throughout the swamps. It is only evidenced by the empty shells often found floating in the tea-colored water, along with the eggs it fixes on the bark of trees and aquatic plants.

water, creating concentric rings that pick up the muted light —blue, silver, and green. It's quiet except for a slight sloshing sound as you walk carefully through the water, feet feeling for cypress knees or logs that might trip you.

What at first threatens to overwhelm begins to differentiate itself. Cypress and red maple trees rise toward the sun. A few huge cypress live on, as old as the redwoods, survivors of intensive logging fifty years ago. Stumps remain, too, a testament to human rapaciousness. Mostly the cypress are of the small, pond variety, but they are beautiful in their delicacy and awesome in their strength.

Individual notes begin to find their way through the cacophony. Yellow bladder-wort brings a point of sunshine and joy into the shadow where tree trunk and water meet, enthralling the eye above while silently trapping insects and minnows in its carnivorous underwater bladders. The bright pink of a bromeliad

[LEFT]
The barred owl is more often heard than seen. He is big, widely dispersed, and usually silent during daylight hours. He flies silently on soft, delicately fringed feathers.

[BELOW]
Ten-petaled Sabatia's pink tutu beckons and close, the golden stamen, fertile with pollen, thrust their beauty skyward.

The pond apple looks just like a God-made little green apple, and raccoons, bears, and other animals love it. But if you bite into it, you won't be happy. It is very hard and bitter to the human palate, although Seminole Indians and early white settlers cooked it with sugar cane. The tree, a tropical species, is a favorite rack for orchids and other epiphytes.

Ferns reproduce through the release of spores from tiny sacks called sporangia, which appear on the underside of fertile fronds. Dozens of species of ferns can be found in the wetlands of Southwest Florida.

[LEFT]
A white ibis, cousin of the pharoah's sacred ibis of legend and reality, stretches its wings, a dancer in Alvin Ailey's production of Grace.

[BELOW]
The clamshell orchid is named for its shell-like petals, which contrast with the surrounding greenery. Devoid of a profound fragrance, it must attract any insect willing to pollinate it by using its showy colors alone.

The swamp is subtle, especially in a winter morning fog. The heavy, moist atmosphere mutes and dilutes the full-color spectrum offered by the rising sun. Soon the sun will prevail, but until then, imagination and spirits reign.

stabs your heart with pleasure. Then you hear a rustle. It is a sense rather than a sound of movement. And then, curved beaks aflame, glass blue eyes find yours. White ibis — one, two, maybe six — bob and probe in search of fish and crustaceans. It's spring, and the new cypress needles beg for fondling. They are fawn soft, breathing the newness of life. Concentrating on one small point at a time begins to bring some order to your view. Sweeping vistas are absent unless you look upward, straight past the convergence of tree tips to the great sky.

Orchids are the special treats — eye candy of the swamp. They still exist in profusion in isolated pockets despite

[ABOVE LEFT]
Shadows and fog, dew and mystery, give way to the force of subtropical sunlight.

—

[ABOVE RIGHT]
The close tree trunks of the pond cypress can be quite gloomy just before dawn, but then the clear winter sun bursts through.

—

[LEFT]
Orchids sometimes beg your eyes to find them. The tiny Epidendrum anceps *is often overlooked.*

29

poaching and the effects of water deprivation. They are ephemeral, blooming only to be pollinated and reproduce. The orchid, some say, is the perfect flower, both male and female in form and shaped to perfection by an unseen hand. It is sensual — visually and sometimes olfactorily exciting. Seeing for the first time a cigar orchid — huge, golden, and red-orange, trying to seduce the bees that no longer live here — is sublime. You cannot drink its beauty enough, looking at it against the velvety-gray cypress trunks, through ferns and lianas, close up, then farther away. You become dizzy with pleasure at the natural beauty and perfect form of this gift. But who sees? Very few. She dances alone for a lover who will never visit her again. Without humans, the magnificent orchid would flower and wither and flower and wither again

without ever producing another of its own kind.

In a moment, as a cloud mutes the sun, experience and imagination are melded into an emotional vortex. It would be so easy to pull away from the real world — so easy to be seduced by the green and gray and tawny color harmonies into a timeless zone. The reporter seeks order and misses the essence; the poet is comfortable with chaos but seeks emotional involvement in, for example, the lives of hatchling alligators nearly lost in a forest of duckweed and doomed, mostly, to be some bigger creature's fare.

Sabatia's pink tutu beckons. Close up, the golden stamens, erect and fertile with pollen, thrust their beauty skyward, demanding visual caresses. The light is tangible, tickling pistils and stamens together, adding a flirtatious breeze to wrap both in a gentle sway. Odors are fresh, sometimes sweet, the proverbial spiritual interior of nature. Relax into it all — the sensual and the spiritual— and serenity is as close and tangible here as anywhere on earth.

Where Are We?

"Understanding wetland ecosystems is a major goal
of contemporary ecological science, and is a necessary
prerequisite to land use management philosophies
that will permit them to be a harmonious
interface between humanity and nature."

Cypress Swamps
Katherine Carter Ewel and Howard T. Odum

S wamps are wetlands, but not all wetlands are swamps. The Southwest Florida swamps are not the stagnant stink-holes full of slime and goo and creepy crawlies of your nightmares. To the contrary, they are lush and green — nature's cistern to the aquifers below — cool water against your legs. They are rivers, perhaps, but are so heavily wooded that water flow is often imperceptible. These swamps are like the mystical rainforests of shampoo advertisements. Yes, air plants hang from every tree, and ferns adorn nearly every stump and fallen log. Orchids, too, grace this kind of garden, but sometimes they beg your eyes to find them. Of course, there are no mountains or hills. The surface is very flat — or so it seems to the human eye. To the vegetation, however, a few inches make a lot of difference. Subtle differences in the topo-graphy create subtle differences in the water levels and how long there is stand-ing water each year — the hydroperiod.

[LEFT]
As the strands feed water to the coastland, the cypress and mixed swamp tree species give way to more open marsh habitat.

[BOTTOM LEFT]
Once common throughout the swamps, vanilla orchids are now rarely found, let alone a huge, sinuous vine like this with such a magnificent bloom.

[OVERLEAF]
During the rainy season (May through October), it is hard to believe that there could ever be a drought, or understand that the ground will get dusty nearly every winter. The swamp has a cyclic ebb and flow, balanced and divinely processed.

[INSET]
Pond cypress. Sweeping vistas are absent unless one looks up, straight up, past the convergence of tree tips to the great sky, bluest in winter.

Water absorbs the sun's energy by day and slowly releases its heat by night. This keeps flooded swamps several degrees warmer, enough to ward off a winter freeze — a rarity, but devastating to the orchids that may take a decade to produce a first bloom. Everything is connected to water. Every living thing in the swamp depends on the amount and timing of the water.

The Big Cypress Basin encompasses much of Southwest Florida's swamp system, and the flat topography translates into expansive strands or slightly deeper sloughs — long, lazy, seasonal flow-ways in shallow, elongated depressions in the bedrock, mostly moving north to south toward the Gulf of Mexico. The Fakahatchee Strand, the largest cypress–mixed-hardwood strand

[ABOVE]

The great egret epitomizes Florida. In South Florida, wading bird populations have decreased by 90 percent, mainly because of wetland drainage and development.

[RIGHT]

The diminutive diformes was once common throughout the swamps, but now it is rare. The entire plant is the size of the palm of your hand. It clings to a cypress stem by its epiphytic roots.

in the United States, is up to two miles wide and meanders twenty miles to where it gives way to coastal marshes and mangroves of the Ten Thousand Island Archipelago.

You can walk in it. In fact, to approach the most beautiful parts of the strand where *Guzmania* thrive and the rarest of orchids are found, you have to. Don't fret — even in the wet season the average depth is only a couple of feet. This doesn't mean you can't drop in a hole, masked under the dark tannin waters, and wish you had a snorkel. Sloughs — the deeper areas by only a foot or two — are not necessarily continuous within the swamp. Here the more constant presence of water throughout the year limits the undergrowth that retards flows and hastens its search for the Gulf. The big bald cypress is the dominant canopy species. It is a regal tree that needs the water, not to germinate new seedlings (it takes the dry season for this) but instead to reduce competition from other trees such as oaks and pines on upland sites that cannot tolerate so much water. Red maple, pop ash, willow, and swamp bay also flourish lower down, filling in the gaps and almost willfully trying to entangle you. "Stay here a while," they seem to plead, "Appreciate me, too."

Cypress domes, small forests, perhaps only a few acres in size, are different from strands. Domes typically occur on depressions or sinks in the landscape, often formed where poor surface drainage has encouraged dissolution of the carbonate bedrock that dominates the Big Cypress, to form something of a pool. Water is powerful when time is its partner. Domes often have a pond in their center where scads of wading birds — often several thousand — commune nightly. Perhaps they congregate to tell of the day's catch of fish or bugs, or tell lies, or sometimes to nest. Domes are

[ABOVE]
Red blanket lichen grows symbiotically as a combination of both alga and fungus.

[LEFT]
The resurrection fern attaches itself to the tops of the larger limbs of old live oaks as a rule, but can be found on other trees and even on the ground. During the dry months it turns brown, but when it rains, it becomes engorged with water and looks "resurrected."

No rivers feed the Big Cypress Basin. Rain alone fills the strands and sloughs and nourishes the cypress and all its neighbors.

The tiny green tree frog is hard to see but easy to hear singing in the waterside vegetation at night.

named for their shape as you view them from a distance. Their canopy is composed almost entirely of cypress. The trees on the periphery are smaller, with only inches of soil over rock, stunting their way upward. The trees grow faster and higher near the center, where many seasons of falling leaves and other detritus escape oxidation under water for so much of the year. Time and water, together again, form organic peat. As deep as six feet, this organic black humus, burnable when dried, is reserved as food to renourish the very trees that created it.

Whether you are among the strands or the domes, the swamps reveal many colors. But the cast of green over your scene is undeniable. The rest of the spectrum is more elusive. Beneath the canopy, the midstory pop-ash and pond apple grow, and saplings of cypress and maple battle for position when a hurricane or some other fate of one of the giants opens a hole to the sky. Fallen logs and shallow areas explode with perhaps two dozen species of ferns. More than a dozen species of *Tillandsia*, commonly called air plants, paint green up the trunks of nearly every tree, wherever their roots can embrace. Shade is plentiful near the swampy floor. Sunlight is scarce, but the lushness belies every plant's struggle to create energy — to photosynthesize — from what little sun it receives. The green color is nature's power plant at full production. Even the orchids, this nation's most diverse collection, have chlorophyll coursing through their roots as they cling to the bark of trees exposed to filtered light from above. Duckweed, made up of several species of minute plants, some as small as a pinhead, grows quickly across the seasonal waters of swamps before the dry of March or April turns it back to the humus from which it sprang. It covers acres of the dark waters with a smooth, manicured, golf-course look.

Swamps are also an open forest, a transitional habitat between the heavy woods and the prairie. Here are the dwarf or hatrack cypress — gangly, stunted, and large buttressed. They are small but old, bizarre in shape and seemingly made for elves. Their spells keep the trees low enough so they can hang their hats on the limbs while they dance their jigs in the moonlight — or so it seems. Ignore the scientists who say they are this way because of nutrient-poor sand and marl over ancient bedrock. Those scientists

say that the roots struggle against geology and that this is why they rarely reach thirty feet in height. The elves can dance dry for two-thirds of the year. The other third, the wet season, exists so the Old Man's Beard can grow and the *Sabatia* can peek around and greet the grass pinks and the thunderclouds.

The human species came to this land as consumers. Confronted with the swamp, to be hated and feared or seen as an opportunity, it has been harvested, drained, and burned. Homes and livings have been built from it. During the 1940s and ending in 1953, nearly all the ancient virgin cypress trees in South Florida were cut for timber. Only those at Audubon's Corkscrew Sanctuary and Big Cypress Bend — fewer than 12,000 acres among more than two million — were saved. Today, cypress will take hundreds of years to reach the majesty of their ancestors.

We are educated by scientists and conservationists who have illuminated the many natural processes that are governed by the swamp. Their value is

The profuse growth of Tillandsia *throughout the swamp adds to the rich diversity of life here.*

in what they hold for us but not in their demise. Swamps have proven to be centers for wildlife production, water conservation, and flood protection. They will again give us timber. If we insist, swamps will give their all, but then there will be nothing for tomorrow's generation. What they will not give away is all their secrets. We must search out and enumerate still further their invaluable gifts to us so we will never forget why we must keep them for humanity. If nothing else, we must resolve that their beauty and mystery are ample gifts to ourselves.

The Sentinel

—

"The cats are predators, . . . hunters, stalkers, climbers, takers. There is no value judgment in that.
To judge the cat because it takes its prey is to question the wisdom of the bluebird and to despise
the purple martin for its meal of mosquitoes near our picnic site. You cannot judge a fragment of nature,
you cannot use wisdom like surgery to extract and judge, lift free of its matrix any small shard of life,
and nod in approval or disapproval. That is not only arrogant but, like all arrogance, foolish."

The Big Cats: The Paintings of Guy Coheleach
Roger Caras, former president, ASPCA

No other animal in South Florida symbolizes the wild in wilderness more than the Florida panther. Panthers are unseen but for the tracks they leave, their scent scrapes on trails, and scratchings on the bark of trees where they have sharpened their claws. The few people who have seen them have had only fleeting glimpses or have treed them with trained dogs for bounty, sport, or science. The Florida panther is a cougar, one subspecies of thirty known to inhabit nearly every ecosystem from as far north as Canada to the southern end of Argentina. Only one other mammal in this hemisphere has a greater range than the cougar — the human being.

Scientifically hailed as a top-of-the-food-chain predator like us, they also hold the dubious designation of "endangered species," a title earned by their plight — the likelihood of extinction if people do not successfully intervene. Officially, only about sixty adults comprise

[ABOVE]

Metaphorically speaking, the Florida panther is up a tree. Only the limb it stands on keeps it from falling into extinction.

—

[OVERLEAF]

In the early 1990s, state and federal agencies established a captive breeding program with an eye toward releasing panthers to the wild. No one had thought how a kitten could be taught to provide for itself in the wild, or where land would be available for additional panthers to live. The program has been abandoned.

—

[INSET]

Feline eyes are watching us. This is Florida panther land, and we are but visitors.

the entire breeding population of Florida panthers, all but one residing south of Lake Okeechobee. Each lives where the human population is minimal, game is ample, and the wilderness supplies all the other amenities.

Who really knows how many panthers are out there? Each male's territory is vast — up to 200 square miles. A female's territory is about half that, albeit still huge. Panthers often trek ten to twenty miles a night and eat an entire deer or wild hog in a week. No salad please. They are strict carnivores. Although not the fastest predator, they are, nevertheless, quite fast, springing to chase down unsuspecting prey at speeds of thirty-five miles per hour. Their tail, nearly one-third of their

[ABOVE]

This helpless fur ball, sightless until about two weeks of age, depends on its mother for all its needs. Its dark spots actually afford camouflage amid the forest undergrowth, concealing the baby panther.

[LEFT]

Adult female Florida panthers associate only with males when their instinctive need to procreate outweighs their inherent disdain for the opposite sex. The two will mate repeatedly for several days to a week; afterward, their paths may never cross again.

[LEFT]
No animal better represents what is wild about wilderness than the Florida panther.

[BELOW]
Like any cat, a panther needs a good scratching post. Panthers have exfoliated the bark on this tree as high as six feet up on the trunk.

tawny six- to seven-foot length, provides balance and steerage as they sprint through the woods after almost equally alert and maneuverable quarry.

There is no elegant courtship. A female will breed with a male and the two will go their separate ways. Ninety-three days hence she will drop her kittens, one to four, in dense undergrowth of palmettos or similar vegetation. Life is hard for the offspring. If they are lucky, they will survive beyond mother's suckle and instructions for hunting, and will leave her after a year or so to range in solitude just as she has done, and her kin before her. We wonder if, years later, two brothers meeting on a trail will know each other and respect one another's right to survive. Perhaps not — fights between male panthers over territory can be brutal. The loser is often left severely wounded or dead. To the victor go the spoils — the right to breed with every available female in that territory, and hence the right to pass on his genes.

Panthers have been accused of living only for eating and procreating. To us, they are sentinels of the swamp, not forbidding our entrance but, rather, declaring it and identifying what is still good and right with this piece of the earth — a guidepost that declares we are in the wilds of South Florida. When their home is no longer able to support them, they will become extinct. They are an umbrella species, protecting all elements of life in their environment, and it is the task of humans to protect the panther. Panthers are charismatic, loved for their beauty, strength, and mystery. They yield to nothing but us, their stewards.

[ABOVE]
As prey for panthers, deer cannot always escape their grasp. Still, they run every race as if it were their last.

[RIGHT]
The Florida panther is the only member of the cougar family now naturally living east of the Mississippi River.

Keystones and Harbingers: Gators and Skeeters

"Never insult an alligator
until after you've crossed the river."

Oriental proverb

Every newcomer or visitor seems to want to see an alligator. It's hard to understand why, because for the most part — although they look sinister — alligators just lie there basking in the sun or slowly cruising along, with their ears, nostrils, and eyes on the same plane, barely clearing the surface. Is it our fascination with potential danger? Few wild alligators present danger, especially in the cool months when, because they are reptiles and do not maintain body heat internally, gators eat infrequently. Wild swamp gators — those not habituated to human presence — fear human approach and slide away unless a female is protecting a nest or her young hatchlings, or unless an amorous male mistakes a biologist or photographer walking in the swamp for a potential mate. Then, look out!

Does a certain part of our brain, our so-called "reptilian brain," identify, from somewhere deep, with the pure instinct driving these animals? Maybe. They are, after all, hugely successful

[ABOVE]

Menacing and hissing, this gator did not back off.

—

[OVERLEAF]

The grin does not match Kipling's crocodile, because the alligator's snout is broader and its teeth appear less menacing. Both crocodiles and alligators can, however, snap their jaws shut with enormous force.

—

[INSET]

A tiny pink Sabatia is a resting place for a huge swamp mosquito enjoying the hot, heavy atmosphere.

actors on the world's stage, brothers to dinosaurs and unchanged for eons. Are we humans, evolving socially at breakneck speed, secretly envious of the gators' unchallenged niche — their timelessness? Maybe.

But people have challenged their niche. In the nineteenth and twentieth centuries, we killed alligators for fashion — shoes and bags — and also killed alligators as vermin. We understood so little. Yet the health of the swamp depends on alligators. As the water recedes in late winter and spring, alligators keep areas of swamp clear and deep — the so-called alligator holes — thereby saving fish and their predators. A keystone species, they, in fact, keep the entire swamp food chain intact.

Yes, unlike most reptiles, the mother will protect her babies. She may actually stay with her young for a year or more or until her next clutch hatches, but after that, they are on their own to eat or be eaten. Hence, biologists count

Florida's alligator population, trying to determine, among other things, their survival rates and where and how they thrive. All that information helps to maintain a robust ecosystem.

Alligators court in May. The males literally vibrate and splash the water around them, with deep, resonating bellows. Around the end of June, the female lays her eggs in a mound that she will protect from raccoons and other animals that would enjoy a meal of eggs. When the babies begin to hatch, they emit a little peep and the mother scrapes away the earth to help them emerge. If the distance is too far for babies to just tumble into the water, the mother will carry them to water in her mouth or on her snout. She will protect the pod, because baby alligators are on everyone's menu. The babies, in turn, will eat any frog, insect, or small fish they can catch.

We were walking in the swamp one sunny day, camera gear in a small pirogue, when we came upon seventeen

Alligators, being ectotherms that have to raise their body heat by sunning, haul up on banks, especially during the cooler months. Alligators knew dinosaurs and survived them, but still, with shiny scales and ferocious claws, they look as if they belong to the Pleistocene Era.

Baby alligators are not good swimmers and are vulnerable to predators,
so they often climb onto water lettuce to bask and better view their surroundings.
Before he grows dark and evil-looking, even a baby alligator can look cute.

hatchlings warming themselves on a log, peeping. Uh-oh — where's the mother? Wordlessly, we both made a dive for the tiny boat, nearly capsizing it, but we continued to photograph the babies until they saw us and slid into the water. We never did see the mother.

Weeks later, and then a year later, we were back catching, counting, weighing, measuring, sexing, and notching the hatchlings' scutes for science and releasing them within a minute or two. How would we not get eaten in the process? We sneaked up at night, when their eyes reflected the flashlight, nabbed them with Pilstrom tongs — which are more commonly used on snakes — then gently held their mouths shut, putting a rubber band around their snouts for insurance. A rubber band actually worked for the small ones and a bit of duct tape sufficed for the larger ones. Okay — we use rope for the really big ones. That's the secret to the final trick of gator wrestling: Alligators can slam their mouths shut

Researchers work at night to catch, count, sex, weigh, measure, and tag alligators.

with awesome power, but it takes only minimal pressure to keep them shut. They have almost no mouth-opening muscles!

Alligators are back from the brink of extinction now, thanks to laws and vigorous husbandry. Noah may have saved them first — or maybe they merely floated beneath the ark — but we have saved them again for our children and for our children's children, and for the perpetuation of the swamp itself. This is a legacy of hope for the unsung wilderness of the Florida swamps.

Miles Scofield and other old-timers say that after the first 500 bites every season, your body becomes immune to the blood thinner that mosquitoes inject and therefore you aren't bothered as much. Mosquitoes, they say, are the reason this region was not inhabited sooner.

Mosquitoes select their victims carefully. Larry is mosquito ambrosia; Connie's blood is liver and broccoli. Why? Researchers at the University of Florida think the excretion of cholesterol and B vitamins, which mosquitoes need but cannot make on their own, might be what stimulates a bite.

Though their bites are selective, mosquitoes will buzz everyone and drive them insane. Mosquitoes are attracted by any hot, carbon-dioxide-laden breath that advertises warm blood on the menu. Sweat is an

[ABOVE LEFT]

Small mosquito fish thrive on mosquito larvae. Many would say they are neither plentiful nor big enough, but without them the balance would swing even more in favor of the insects. Although mosquito fish are only about an inch or two in length, they can multiply rapidly in warm, wet weather. Their young, delivered live, grow rapidly.

—

[ABOVE RIGHT]

Too young to fulfill its role in the swamp, this yearling alligator will someday have dominion.

—

[LEFT]

The royal palm has been fashioned into a symbol of an instant upscale Florida resort, but she is, in fact, a daughter of the swamp, our swamp, a native. Looking down through the dark water into her upper branches, it would be so easy to pull away from the real world, to be seduced by the green and gray and tawny color harmonies into a timeless zone.

[ABOVE]
*The common cooter, like its cousin
the redbelly, basks during the day,
especially in the cool months.*

—

[ABOVE LEFT]
*Wood storks continually add large sticks to their
flimsy platform nests, preferably at the top of
large cypress standing in water. The greatest
concentration of nesting wood storks can
be found at the Corkscrew Sanctuary.*

—

[LEFT]
*Amongst tall sawgrass and ferns, a deer
exercises its senses — hearing, smell,
and vision — to detect friend or foe.*

appetizer, whereas certain heart and blood-pressure medicines and scented bath products are like candy.

Mosquito larvae and pupae can live wherever there is standing water. The larvae feed constantly on pieces of dead plants and animals, and, in turn, they are eaten by mosquito fish *(Gambusia)*. In fact, *Gambusia* often are used for mosquito control.

More commonly, Florida mosquito-control agencies use larvicides and insecticides along with natural predators and parasites to control mosquitoes in urban areas, but the swamp is left in its natural state. (Translation: Zillions of mosquitoes, even after *Gambusia* and other predators have had their fill.)

Mosquitoes may be the deadliest animals on earth. Everyone knows that they carry disease, including malaria, though in the United States the risk is slight. AIDS cannot be transmitted by

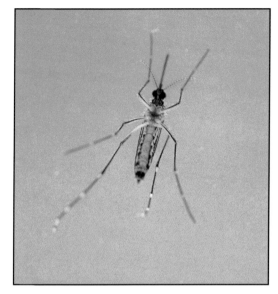

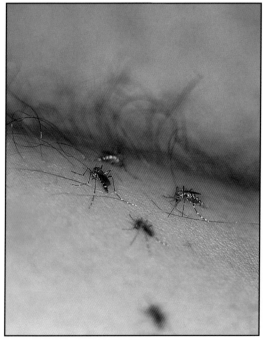

mosquitoes, but encephalitis can be. The St. Louis strain can be fatal, but fortunately, it's rare.

Mosquitoes like cocktails, sunsets, night activities, and sunrises. They are enamored of bare ankles and perfume. So the wise tactic is to cover up at dusk and stay out of the swamp then. That, of course, is when photographers find the magic light. Connie therefore goes into the swamp with Larry — he's her best mosquito repellent.

[ABOVE]
Worse is looking at the underside of a female mosquito preparing for her blood supper. But look closely — she is almost pretty with her golden stripes.

[BELOW]
Female mosquitoes waste no time engorging themselves on human blood.

Vixen

—

"To you, I am nothing more than a fox like a hundred thousand other foxes.
But if you [establish ties with] me, then we shall need each other. . . . To you,
I shall be unique in all the world. . . . You become responsible, forever,
for what you have [established ties with]. You are responsible."

The Little Prince
Antoine de Saint-Exupery

We canoed out to the island, hoping not to find her, but the fox odor was strong. Tying up to a cypress knee, we stepped onto the springy humus and over to the base of the huge cypress, one of the few old ones still standing. In a strange turnabout, its embracing fig had saved it from the logger's saw a half-century ago. Loggers had shunned the trunks hollowed by the parasitic fig's consumption. Now the tree faced a slower demise from the lianas' slow strangle.

The vixen was not in the tree where she had been when we left her a week before, nor was she hiding in the dead leaves of the saw palmetto that had fallen to the ground. The pieces of cooked turkey we had left were gone, as were all the cashews, but a few Brazil nuts remained where we had left them for her. She had defecated, and urinated to leave her scent. We poked around on the little island — actually, just the base of the cypress and debris built up over

[ABOVE]

The gray fox, also called the swamp fox, survives by its wits. It eats small mammals but also partakes of amphibians, insects, fruits, birds, and eggs.

—

[OVERLEAF]

The huge cypress, a favorite of the fox, was one of the few old ones still standing. Its embracing fig had saved it from the logger's saw, only to meet a slower demise from the lianas' strangle.

—

[INSET]

A gray fox hidden in wax myrtle and swamp fern, watching for intruders.

58

the course of the years, along with palms, wax myrtle, and swamp fern taking hold on top — but we did not find her.

She had apparently swum the thirty feet or so to firmer ground, to a bigger piece of dry earth — or so we hoped. Relieved, we returned to the canoe and went looking for orchids in bloom or anything else the swamp would offer up to our eyes and our cameras. The light was too beautiful to waste and tempted us deeper and deeper into the labyrinth. We paddled almost silently, but it was not quiet. It was spring and the migratory songbirds were announcing to leaves, to butterflies, and to all creation that they were bursting with hormones, seeking a mate, and heading north. Occasionally the harsh squeak of a swamp turkey, the anhinga, could be heard from a low branch as it opened its oilless and saturated wings to dry.

We found no orchids in bloom today, so we worked our way back toward the old fishing cabin, a relic of days when humans came only to take trees, turkeys, deer, pigs, and especially fish.

[ABOVE LEFT]
A yellow rat snake.

—

[BELOW LEFT]
The vixen was wary but not afraid of us. She watched us. She is quite capable of killing her own food, but whether she thrives in the swamp or dies of starvation, her legacy is a symbol of our intervention.

—

[RIGHT]
Virginia creeper feels its way up the trunk of a tree.

"Let's check the island one more time," I suggested, but Larry had already steered us in that direction. From the canoe, I saw her, maybe twenty feet up the tree, curled snugly in the crotch of the first fat limb, watching us. Gray foxes are the only American canid with true climbing ability, not so much hugging the tree but, rather, scrambling from toehold to toehold to find a horizontal resting or foraging place. That was easy in this giant. Liana ledges were sturdy enough for a fox, and the limbs, starting fairly low, were broad and flat and tempting even for a human nap.

We both sighed. Now we would have to catch her, upset her, and take her across for release on drier ground. Larry let me off with my camera gear, then paddled back for the cages. I climbed up slowly so as not to scare her and shot several closeups from about four feet. She was curious but not disturbed.

This young vixen had been the subject of a court case just two weeks earlier. She had been found as a kit, fed, and kept until she grew to maturity. The state officials had heard that someone had a wild fox without a permit and they confiscated her, then deposited her at the Wildlife Rehabilitation Center in the Conservancy of Southwest Florida. There she stayed while the case proceeded. Because the vixen would be released after the court heard the matter, rehabilitation volunteers and interns gave her only live mice, thereby teaching her how to hunt and kill for herself. By the time we took her to the swamp, she was capable of feeding herself, and by not being handled for many weeks, she had become more suspicious of humans. This was good.

We released her on the island so we could monitor her progress more carefully, hoping she would swim away whenever she was ready. Swamp foxes are known to be excellent swimmers. A week had gone by and she had not left the island. Although swamp foxes eat a wide variety of foods, including mice, birds, reptiles, insects, and even berries and fruit, eventually the meager food sources on the island would be exhausted. We had to move her.

I could still get my hand on her. I wanted to scruff her — hold her close while she hung like a kit being carried by its mother — as I scrambled down the tree. I then would put her into the large traveling cage, then get the cage into the canoe so we could paddle back to the cabin and release her. Larry, however, did not want to risk it. He brought a Have-a-Heart trap and the chicken pieces I had brought for the fox. He deposited the trap in the most open part of the island, between the base of the tree and the water line. As he came around the tree, he knocked a dead, dry palm leaf to the ground with an almost deafening rattle. Startled, the vixen leaped past me, down the tree, and straight into the fern and myrtle brush. Three feet from us, she was completely hidden. Now we knew where she had been when we had looked earlier.

Patience, now, would bring her out — or chicken. Chicken did the trick. We threw her some so she could taste it, and left a trail to the trap and into it —

a big, succulent morsel well inside. Then we retreated to the canoe to watch from the water. It took a while, but she was obviously hungry and eventually she was caught inside. We moved fast now to prevent her from becoming unduly stressed.

Her new home was a rectangular area, somewhat open in the center but surrounded by brush. It was dry, even when the lower ground was immersed. We let her out — only about twenty minutes had elapsed — and left the rest of the chicken and some water for her. The next day she was seen, still happy but wary. She has not been seen since, nor has there been any sign of her demise. We have seen prints, much like those of a domestic cat but with the nonretractile claws showing in the soft, wet earth, and we have smelled the pungent odor of fox. Gray foxes have been seen regularly nearby, and we hope she, now a free young vixen, has joined her kin.

Stealthy and quiet, the gray fox is rarely seen, but occasionally one can be spotted in a large tree in which it forages and hides. The gray fox is the only American canid with true climbing ability, allowing it to go high up into a tree for protection.

Mystery of the Bee-Swarm Orchid

"For . . . I am a bee, . . .
And I wish to state that I'll *always* mate
With whatever drone I encounter."

Song of the Queen Bee
E.B. White

Year after year they would don their showiest colors — gold and red, ruffles and polka-dots — folds and folds of rich, bumblebee colors. Year after year they would tire of waiting for a mate — a pollinator — to perpetuate their kind. They would wither, draw back into themselves, shed their party clothes, and, barren yet again, settle back to everyday attire with their large, clublike pseudobulbs and stout, strap-shaped leaves for another year. These long pseudobulbs give this orchid its other common name — the cigar orchid.

Where is the pollinator to carry the species onward? The name "bee-swarm orchid" arose because a swarm of bees is like the profusion of resplendent blossoms on long, inflorescence arising from its base. Certainly the size of the flower — about two inches — and color are beelike. Where is Sherlock Holmes when we need him? And why do the plants not reproduce? How can a drone resist?

Once common in South Florida, the cigar orchid, *Cyrtopodium punctatum*, is now rarely seen. Only a handful are known. Epiphytic like many *Orchidae*, they usually are found attached to high tree limbs, where light is abundant and choking weeds are absent.

Orchids are not parasitic. *Cyrtopodium* wants high humidity, a distinct dry season so the plant can gather energy for its exuberant display, and sufficient standing water during the coolest months to protect it from the coldest nights. Therein lies another dilemma: Without enough water — without the swamp itself — *Cyrtopodium* would be vulnerable to a freeze. The water below insulates and protects. It absorbs the warmth of the sun during the day and releases that warmth at night. One drought year combined with a winter freeze, and even these few remaining individuals would be gone forever.

The orchid family, perhaps the largest family of flowering plants, wild and hybridized, terrestrial and epiphytic, encompasses plants a fraction of an inch high to huge specimens with flower stalks fifteen feet long.

All orchid flowers have three sepals, or outer whorls, and three petals, or inner whorls. One petal, the labellum, is larger and showier than the others. On *Cyrtopodium*, as on most orchids, it is the lowest segment. Projecting from the center of the flower — with a pale yellow fading to a soft, cool green center — is the fleshy fusion of the staminate (male part) and the pistillate (female part) of the reproductive organs, unique to all orchids. At the top of the column is the anther. Its pollen grains, the pollinia, perch on the apex. Immediately below are the female parts, the stigma, a sticky indented surface on which the pollinia are deposited.

Unfortunately, the pollen cannot simply fall from the anther right into the stigma and "Elementary, Watson," we have pollination. Propagation of this orchid is not easy. This highly intricate plant

[LEFT]

Unless humans intervene, the cigar orchid would be relegated to a tenuous existence without offspring, eventually to disappear from Earth forever. Enter the biologists, who take tiny forceps to pluck the pollinia, a single waxy cluster of pollen half the size of a pin head, and implant it in the pistil of another plant in the hope of sustaining the species.

—

[BELOW]

Projecting from the center of the flower, a paler yellow fading to a soft, cool green center, is the fleshy fusion of the staminate (male part) and the pistillate (female part) of the reproductive organs, unique to all orchids.

requires cross-pollination. Insects are the usual pollinators. The bee-swarm orchid looks like a bumblebee in order to seduce a bumblebee. Like its kin, *Cyrtopodium*'s flowers are intricately constructed so the male bee is tricked into entering the blossom, thinking — if bees think, especially at a time like this — it is copulating with another bee. Too late, it finds that is not the case and backs out of the flower, picking up the pollinia, like a Post-It Note, on the back of its head. Again, the same bee is tricked by a flower on another plant, this time leaving its Post-It Note pollinia attached to the stigma of a flower on that plant.

Cyrtopodium blooms in vain, however, because her suitors — *Euglossa hemichlora* and *Centris versicolor* — are gone. They disappeared from the swamp years ago, perhaps exterminated by pesticides used on nearby cultivated fields. Pseudo-copulation is over for the *Cyrtopodium* unless, of course, another pseudomate can take pollen from one plant to another.

Unless humans intervene, this orchid is relegated to a tenuous existence without offspring, eventually to disappear from Earth forever. Enter the biologists. Among individual plants in the swamp on the Florida Panther National Wildlife Refuge, scattered among stunted cypress, biologists now take tiny forceps to pluck the pollinia, a single waxy cluster of pollen half the size of a pin head, and implant it in the pistil of another plant a mile or so away.

Its rarity and beauty might be its demise. Besides freezes and an absence of suitors, this showy orchid faces yet another threat — that of kidnapping. Some orchid lovers are notoriously obsessed with owning and controlling the rarest, most beautiful orchids, and *Cyrtopodium* fits that bill. Others who know the swamp well may be induced, because big money is paid for such exotic orchids, to poach the handful remaining. Thus, their location is kept secret. That in itself is a tragedy.

Who Flies There?

"Wild spirit called to wild spirit, and she
seemed to be flying with the great bird,
soaring with it in the evening sky . . ."

The Snow Goose
Paul Gallico

We all want to fly — we earthbound creatures looking up to the sky to escape the mundane. Our poetry and our fanciful longings so often ride wings. Metaphorically, we rise above our humdrum existence, escape on wisps of air, ride the thermals up into disengagement, and delight in the very thought of flight — Icharus notwithstanding.

Among the shadows of the forest — the twilight zone under sheltering greenery interspersed with sharp stabs of harsh light — lurk and feed the ballerinas. These are the egrets, the herons, the storks, and the ibis, with a supporting cast of rails, anhingas, turkeys, and limpkins. The hawks are the bouncers removing the riffraff, and the vultures comprise the clean-up crew. The choir of songbirds asks but does not receive silence from the elegant and graceful but harshly squawking stars. Still, they bring aural delight even when they are hidden behind the

[LEFT]
The majestic great egret displays elegant breeding plumes in the winter. At the turn of the century, the plumes were used to decorate women's hats, and hunting of the egrets almost extinguished the birds. The Audubon Society was formed to stop this slaughter, and, as a result, the population stabilized. Today, however, loss of habitat is again putting pressure on wading-bird populations.

[BOTTOM LEFT]
Looking at the wood stork with its bare, scaly neck and head, one could easily believe that birds are dinosaurs in disguise. The bright red beak signals this stork's breeding fitness.

[OVERLEAF]
Snowy egrets often defend their right to a piece of fishing space, but at the beginning of the breeding season, they congregate along rich waterways, often in the company of great egrets, ibis, anhingas, and vultures.

[INSET]
The red-shouldered hawk is the most common raptor in the South Florida swamps.

curtain of greenery on this stage. The ebb and flow of migrant snowbirds is the audience, challenged by owls asking who goes there, in this drama on land, in water, and in the air. Critters furred and scaled, vertebrate and invertebrate, traverse the trails and waters of the swamp, but the most enigmatic in origin and habits are feathered. They, like our fancy, live on the wind.

Florida, and the Southwest Florida swamps in particular, glory in enormous diversity and numbers of large and small birds. We love birds, but we have exploited them, hunting plumed egrets nearly to extinction, and drained waterways that are wading birds' pantries. Feathers, their most obvious common feature, are truly marvelous but require high maintenance. They

[ABOVE]

In muted blues, purples, and grays, along with the soft morning greens of the swamp — African violet colors — a little blue heron anticipates a frog or fish for breakfast.

[LEFT]

Nearly fledged green herons wait none-too-patiently against the reeds for breakfast waterside. The chicks are dependent for about thirty-five days, then they fledge and hunt for themselves.

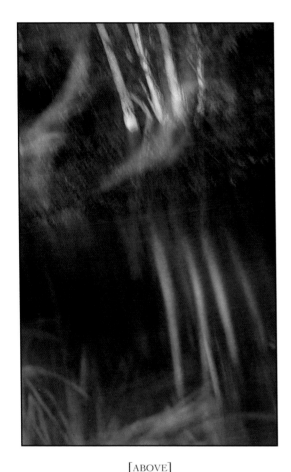

[ABOVE]

*After a few hours in the swamp,
it is easy to believe you have
been transported back eons in time.
The gators, the turtles, the water,
all are timeless. In flight, the
almost-seen images of egrets
transport the soul to another sphere.*

must be preened, bathed, and oiled. In preening, a bird slides its beak over its feathers to connect the individual barbules along the vane that is on either side of a central shaft, or rachis. Oiling is most important in water birds, but all birds must do it. It must rub its beak along the oil (uropygial) gland located on its rump, then spread the oil onto its feathers while preening. Well-preened and lubricated feathers help to water-proof a bird. Clean plumage helps, too, and birds of all sizes bathe to maintain the health of their feathers.

Swamp birds almost always sleep in individual trees or communal roosts. How do they do it? Why don't they fall off of their perches? It's all in the tendons.

As the bird squats, the flexor tendons tighten the toes around the perch; as the bird straightens up, ready to fly, the same tendons relax their pull on the toes, and, with a small push and open wings, the bird lifts off.

Birds eat in several ways — stalking, probing, and scavenging. The green heron, for example, is a patient hunter that waits on low branches to snare a fish. Night herons specialize in crustaceans but don't feed only at night. Snowy egrets, wearing flashy, street-walker-yellow feet, often sprint across ponds, dragging their toes and stirring up fish and attention. Then they dip their beaks and, in a flash, catch their fare. They are party creatures and

[LEFT]

*The purple gallinule prefers dense
vegetation along marsh edges.
Like other rails, its feet are designed
to spread the weight so the bird
can walk on lily pads and other
floating vegetation. It probes for
vegetation but also plucks new leaf
sprouts from marsh willows.*

[ABOVE LEFT]

Rails, limpkins, and bitterns are not uncommon, but they are secretive and conduct their daily routines unnoticed. This limpkin was tolerant of the photographer for almost five minutes.

—

[ABOVE RIGHT]

The sora rail, while not uncommon, is rarely seen. Very shy and secretive, the sora stays well hidden in freshwater marshes and probes for seeds, snails, and other invertebrates. Its coloring is hard to spot.

—

[LEFT]

Never had roseate spoonbills been seen nesting at Corkscrew Sanctuary before the spring of 1999, but nests were indeed there in a mixed colony of wood storks, egrets, and anhinga.

perform in company, each showing off for the next — a precision drill team darting from corner to corner almost on cue.

No bird says "South Florida swamp" like the wood stork. Once numbering in the tens of thousands, stork numbers are now more like 5,000 to 7,000. If they nest at all, they do so almost exclusively in the National Audubon Society's Corkscrew Swamp Sanctuary in the northwestern corner of the Big Cypress Basin. Storks are picky eaters and picky breeders. Everything has to be just right. They need shallow water with just the right concentrations of fish, crayfish, and frogs. They are tactile feeders, groping around with their beaks until they touch potential food. Then — wham! — in $\frac{1}{40}$ of a second, they snap it up. A too-wet year disperses the fish too widely, and storks cannot feed sufficiently to sustain their young. In a too-dry year, there is not enough food. Again, we have messed with Mother Nature, drained and mismanaged the

[ABOVE]
Ready to fly, an immature great blue heron feels the strength of his open wings, knowing he will be king of the wading birds. For now, he is but a prince.

[LEFT]
Red-bellied woodpeckers love the swamp but have adjusted to more urban environments as well, especially when they are fed by humans. Their nests are in deciduous snags or, common in South Florida, in dead palm trunks.

wetlands and disrupted stork feeding and breeding cycles. If they cannot adequately feed themselves and their chicks, the birds abandon their nests, and because they nest in colonies, breeding failure means not just one nest or a few but the entire colony.

If wood storks become extinct, where will babies come from?

If bird beauties decline to rarity, won't our imagination suffer?

Fire in the Garden

—

"Now, sudden fountains of color surged into the night sky, and then, mysteriously, drained away to a small, dim smudge, only to flare up again minutes later. . . . The ashen skeleton of a grass leaf, incinerated by intense heat, settled on the back of my hand. . . . I felt small and threatened. It looked larger, more powerful than I had imagined."

Cry of the Kalahari
Mark and Delia Owens

ire makes its presence known in the swamp as destroyer or savior, defining the swamp along with water from rain and runoff, soil accumulated through geologic processes, years of falling leaves, and time.

With nature at the helm, Fire the Destroyer is rare, perhaps every 100 to 800 years. It must find its way here first through extended drought or at winter's rainless ebb, when vegetation is parched and water is withdrawn beneath the organic peat. Then lightning must arrive — Thor's mighty hand striking a tree, bark splitting down to the ground or exploding into a thousand splinters and burning shards. If the water has not retreated, little is lost but the victim tree. If water is absent, the flames spread, carried first by leaf litter on the ground. The peat soils burn as deep as the water allows, or to where the bedrock starts, perhaps deep enough for the smoldering creep of

[LEFT]
Repeated fires in the swamp have burned away the organic top soil and exposed the support roots of cypress.

[BELOW]
Rain lilies are born of fire. Delicate and tentative, rain lilies emerge after fire, first to welcome the wonders of life transformed.

[OVERLEAF]
The landscape receives the tsunami of fire that sweeps across an expansive prairie into the swamp.

[INSET]
A firefighter takes a moment to view the colors of the awesome power of nature.

78

punky embers to bake the tangled tree roots or scour out their hearts with flame. The cypress and the hardwoods die, monuments to centuries past. If the fire is complete, it leaves a chasm. What was once the swamp — shaded, moist, and thriving green — is black, cluttered, skeletal. It must wait to be renewed by the wind and water, by seeds dropped by passing birds, and by bears and deer.

When the soil has not lost all its moisture, fire acts as Savior, burning the scant surface litter and returning nutrients to the soil — stimulants to good health. The fire must depend on winds to carry the flames, to sweep through a

Prescribed fires reduce fuel loads, recycle nutrients, and stimulate the fresh growth of plants that deer and other wildlife desire.

forest at far less intensity. Perhaps low shrubs here and there will die; so will some smaller cypress. Other hardwoods that are not as tolerant to fire as cypress will succumb to the heat as flames lick their thin bark to death. The canopy opens up a bit and more sunshine reaches the forest floor. New, fresh growth is found here. Succession is stalled. The cypress swamp survives as triumphant — the swamp bays and red maples fought off for now. This is nature's way of thinning the forest — gardening by the Master Gardener.

Historians tell us that the Calusa Indians, on the scene in the first millennium, probably used fire to drive game during hunts. Perhaps they knew the value of fire to regenerate more nutritious forage for deer and other wildlife, as we do today. Since the arrival of Europeans, the frequency of fire has increased. Today, most fires caused by humans occur in the dry season when the frequency of lightning is minimal and fire's effect is more devastating.

Mostly, we humans are helping the Destroyer. We funnel water into canals, thereby shortening the hydroperiod. Land that was saturated up to 325 days per year now holds far less water. Not only are the organic soils disappearing through the process of oxidization, which is retarded when water is present, but the dryness encourages fires

every few years, gradually killing off the cypress and other hydrophytes, removing detritus and organic soil down to bedrock. Nothing can grow. This is not dramatic, but the repeated assault by fire insidiously changes swamp to land devoid of rich epiphytic orchids and ferns, mosses, and lianas — in fact, all indices of a moist, subtropical forest are being exterminated.

Some efforts are being made to help Fire the Savior by rehydrating the land and burning the heavy accumulation of fuels around the swamp, still hydrated, before they build up, fueling the risk of an overheated, out-of-control, destructive fire. New growth — often within a week — entices the return of ungulates, and they, in turn, encourage predators — a vigorous and healthy ecosystem.

A rain lily springs forth after a fire, possibly after twenty years of hiding in the ground.

Good Old Days

—

"If we are to survive as a healthy species, we are obliged to go forward into
this new age with the preservation and restoration of the environment
in the forefront of our thoughts. We must establish an ethic toward
the earth that will penetrate the very soul of our existence."

Visions for the Next Millennium
Clyde Butcher, Photographic Artist

"I've been trying for fifty-five years to get close to a panther." Miles Scofield's clouding hazel eyes twinkled as he explained how he masked his own scent and confused a young adult cat. "I was up on a platform six feet high because I can't climb trees any more." He added wryly that he could no longer pull the bow string back and had not shot a gun in five years. But he wanted to be up there where he once hunted deer and turkey.

Hunting has been a part of the swamp forever. The Calusa Indians, coastal people who populated South Florida for more than 3,000 years, ventured inland for game and medicinal herbs. They were exterminated in the early eighteenth century by Spanish slavery and white man's diseases and were replaced by an amalgam called the Seminoles. They, too, lived lightly off the land, largely because their numbers were few. But white settlers, arriving from Georgia and the Carolinas, were not so kind.

The bounty was for the taking. Hunting of deer and turkey, now tightly regulated, continues in many parts of the swamp. The difference is in numbers — lots more people and lots less game. Scofield says he has never personally known of panther hunting, but as recently as 1998, a panther was shot just east of Naples. Before that, the only organized and extensive panther hunting occurred in the 1940s.

In the 1920s, the Tamiami Trail was cut across the southward flow of water like a dam. Arthur Lee, journalist and self-taught archaeologist, says that the dredges took the heart out of the entire Big Cypress Basin, and that the trail was the end of the Everglades. That changed the ecology of all of South Florida.

Botanist Bob Read laments the resulting loss of precious plant life: "We're having more freezes, and I think it's because of draining. We have lost many orchids and epiphytes, and what used to be cypress and pine is now almost all

[OVERLEAF]
Cloud patterns at Corkscrew Swamp Sanctuary.

———

[INSET]
The white ibis corps de ballet revel in the sun's spotlight.

———

[BELOW]
Feral pigs eat well in South Florida and in turn provide excellent sustenance for panthers.

The most common orchid of the Florida swamps is the butterfly orchid. It clings to a variety of tree species and is tolerant of differing degrees of sunlight. It has a heavenly scent.

cabbage palm. We need a longer hydro-period [to support tropical plants] and more humidity."

Bubba Frank is saddened. The orchids he loves, especially the ghost orchids and butterfly orchids, have nearly disappeared because water drainage has

changed the climate. Something else is happening, too. Two codgers were bragging about how many rare or endangered orchids they had stolen in their day, and where they came from. Even allowing for old men's expanded memories but multiplying their boasts by the many more who did the same, orchids and rare bromeliads were raided ruthlessly, and maybe still are. Perhaps even more significant is the continued theft of rare and endangered snakes and other reptiles.

Scofield, grinning, continued his story. "I just hunkered down and never moved. He zigzagged, following the vanilla scent for five or seven minutes," then put his nose on the second step up the platform. "His nose and my toes were four feet apart. I'm looking right down at the animal! I knew what he was thinking. He didn't know what the hell I was!"

"The ranch always had panthers," and for ten years Scofield worked with the

Florida Fish and Game Commission to understand their needs. Today, he believes they have been studied to death. He commented, "A panther wants cover and food, and we afford cover for them and a lot of food. At one time this ranch raised hogs, but hogs didn't pay. So we just turned them out to be wild pigs. A panther has a little problem catching and eating a deer, but a pig is easy for him. Panthers are doing us a big favor by keeping the pig population down. I know of one breeding female, plus another female is out there. She hasn't had any cubs, but the older female raises two cubs every other year. Males cover a large territory, and they come and go. They love pigs and get fat and healthy on them. And Collier County has the largest panthers in Florida."

Clearly, Scofield is proud of his steward-ship. He and his sons manage the 10,000-acre cattle ranch and citrus groves profitably, and actively encourage wildlife. They burn one-third of their land every year because the tender grass that comes up afterward is good for the game and for the cattle. According to Scofield, a good burn program is vital to a successful operation. He says, "We know how to manage the land, and we can do it a hell of a lot cheaper than any government body. I want to leave my sons this land intact. If the developers and biologists and politicians leave us alone, we'll be all right."

Scofield and many of the other pioneers, mostly ranchers, revere the land and respect it's power. Scofield said that in the early days it took about eight hours to get from Everglades City to the ranch. They would drive up to Sunniland, where they had a swamp buggy parked, then went the rest of the way on the buggy. Now, State Road 29 cuts straight through to Immokalee, with trucks and cars regularly exceeding the speed limit by twenty or thirty miles per hour. Like the Crackers themselves, who take their moniker from the cracking of their long cattle-herding whips,

[LEFT]
This shed at the Scofield Ranch was used in past times for hanging deer, hogs, turkey, or other game.

[RIGHT]
As the sun departs for the day, the blaze of light silhouettes cypress against a clouded sky. The swamps of South Florida survive another day, only to begin the photosynthetic process tomorrow.